13 FACTS THAT PROVE HUMANS DON'T CAUSE GLOBAL WARMING

Stop Blaming Carbon Dioxide For Climate Change

Alan Fensin

Published by:
Burlington Book Division
Burlington National Inc.
Box 841
Mandeville, LA 70470
http://www.13facts.com

ISBN 978-1-57706-656-9
ISBN 1-57706-656-1

Printed in the United States of America.

13 Facts That Prove Humans Don't Cause Global Warming is for everyone who doesn't really understand the complicated climate change controversy. It is short, easy to read and written for the non-scientist. It simplifies everything about global warming and climate change.

The global warming alarmist case is based entirely on misrepresentation. Certain scientists nowadays think nothing of skewing their scientific research for political purposes. They say carbon dioxide causes Earth to get warmer. But the truth is getting warmer causes increased carbon dioxide. They say humans cause the Earth to get warmer. But the truth is Earth's climate has been changing long before humans lived on Earth. They say carbon dioxide is the major green house gas. But the truth is water vapor is the major green house gas and in comparison human contribution to carbon dioxide is trivial.

Inside you'll find spectacular proof that humans are not responsible for global warming. Mother Nature controls our weather and always has.

You will journey through the politics and science of global warming with numerous facts including quotes from top scientists. With this roadmap, you will understand the climate change arguments. This is the ultimate book on the man-made global warming myth.

Table of Contents

Overview of the Book

Man-made global warming is not a fact! It is only a theory. And this book will show that it is not even a good theory.

Man-made global warming is also the single greatest scientific misinformation since the seventeenth century. That was when the famous scientist Galileo denied that the Sun revolved around Earth and said that Earth revolved around the Sun. Galileo was arrested and threatened with torture for being a denier. The government insisted that the Sun went around the Earth. Galileo spent the rest of his life under house arrest and was not allowed to spread his lies.

> "In questions of science, the authority of a thousand is not worth the humble reasoning of a single individual". - Galileo Galilei

Today, in an eerie repeat of Galileo's history, scientists say that humans are responsible for climate change. And, as with Galileo, the political powers have a theory that says humans and some small trace gas called carbon dioxide are responsible for controlling our climate.

The mainstream media, together with greedy lobbyists, eagerly promote climate change scare tactics. They disregard the real facts and repeat the same old lies over and over. The result is more big government spending and many billions of dollars going into the promoter's pockets.

> "It is easier to silence scientific dissent by utilizing the politics of personal destruction, than to actually debate them on the merits of their arguments." - Mike Thompson, Chief Meteorologist

No one has ever been able to prove that humans cause global warming. The only historical facts are that global warming and climate change are completely normal. They are caused by natural physical occurrences and not by humans.

The global warming alarmists are wasting many billions of our dollars every year, and they want much more. In fact they want trillions. But they have absolutely no evidence that humans are causing global warming. All the alarmists have is a theory they are unable to prove. And their theory is wrong. This book will present thirteen facts that prove they are wrong.

> "The facts, such as we can observe and calculate them, do not support the idea of man-made global warming. Natural processes completely eclipse anything that man can accomplish." - William Hunt, research scientist National Oceanic and Atmospheric Administration

Man-made climate change has been one of the more complicated and polarizing issues of the past several years. People seem to think that the study of Earth's climate means you are either a meteorologist or a climatologist. But that is a very limited view. Man-made climate change is a complicated subject. It covers scientific specialties in solar science, computer modeling, space science, environmental science, geology, cosmology, physics, biology, meteorology, astronomy, ocean science, and more.

So the study of Earth's climate requires knowledge or input from a wide variety of scientific specialties. This makes it even more difficult for the non-scientist to know the right thing to do.

Political groups meddling in climate change science makes it even more complicated. And add to that the fact that misleading and downright wrong statements are purposely repeated over and over.

"Warming fears are the worst scientific scandal in history. When people come to know what the truth is, they will feel deceived by science and scientists." - Dr. Kiminori Itoh, environmental physical chemist.

Not too long ago we were told to fear the evils of man-made global warming. But there has been no increase in the average global temperature since 1998. So in order not to appear foolish, the name global warming was changed and now it's climate change.

Today we are told that practically any extreme climate event that occurs happens because humans and man-made carbon dioxide caused it.

"Climate is changing, and climate has always changed. The hoax is that there are some people who are so arrogant to think that they are so powerful they can change climate. Man can't change climate." - James Inhofe U.S. Senate's chairman of the Environment and Public Works Committee

The mainstream news media typically portrays man-made global warming as recognized scientific fact. Any books, scientific papers, or prominent people who disagree with the global change alarmists face attacks.

If the authors work for the government or are involved in government contracts they face the loss of jobs and money. Scientists who disagree with the alarmist view are ruthlessly attacked and have lost their research grants and funding. This has caused others to understandably remain silent.

> "Belief in climate models is compared to ancient astrology...I believe the man-made effect for climate change is still only one of the hypotheses to explain the variability of climate." - Dr. Kanya Kusano, Physicist, program director of the Japan Agency for Marine-Earth Science

The public's lack of scientific understanding, especially with regard to a complicated issue such as climate science, has been exploited. Global warming skeptics have been ridiculed and called deniers. Climate change now seems to be more about political science and no longer about real science. So the question is "should you trust the climate alarmists' hype or should you trust the scientific method that tests against the real world and says there always was climate change?" The scientific method has always been the best tool for understanding how the world works.

> "You know as well as I, the 'global warming scare' is being used as a political tool to increase government control over American lives, incomes and decision making." - Jack Schmitt, geology scientist and U.S. astronaut

People with very little scientific knowledge, but strong political views, attack scientists whose views differ from theirs. Not only do they ridicule them, but they also attempt to damage their character, their careers, and their very lives.

"Unfortunately, Climate Science has become Political Science...It is tragic that some perhaps well-meaning but politically motivated scientists who should know better have whipped up a global frenzy about a phenomena which is statistically questionable at best." - Dr. Robert Austin, Physicist Princeton University

The only permanent thing is change. Climate change is real and can either do some very good things or some very undesirable things on Earth. But Mother Nature causes climate change and was doing it long long before humans even existed. Variations in climate are not the result of human activity. The best evidence about climate change comes from the past, and our past says that there has always been climate change.

This book was designed to simplify the various climate change proofs against man-made global warming and make them accessible to the non-scientist. Towards that end, I also wanted to make this book easy and short enough so that it would be accessible to anyone and people would not be afraid to read it.

"There is no climate crisis. The ocean is not rising significantly. The polar ice is increasing, not melting away...I have studied this topic seriously for years. It has become a political and environmental agenda item, but the science is not valid." - John Coleman, Meteorologist, co-founded The Weather Channel

First Fact

Long before humans existed, there was dramatic climate change.

What is an Ice Age?

Do you believe we are currently in an ice age? Most people without education in Earth sciences say no. But in fact we are now very definitely in an ice age.

The definition of an ice age is the presence of permanent ice sheets in both the Northern and Southern Hemispheres. And our current ice age certainly fulfills that definition. So we are now in an ice age.

> "You know, to think that we could affect weather all that much is pretty arrogant. Mother Nature is so big, the world is so big, the oceans are so big – I think we're going to die from a lack of fresh water or we're going to die from ocean acidification before we die from global warming." - Chad Myers, meteorologist

How long do they last?

We are in the fifth ice age that science knows about, named the "Pleistocene glaciation" ice age. We have been in this ice age for slightly more than two and a half million years.

The previous four ice ages lasted considerably longer than our current ice age. The odds are that our current ice age will last for a very long time. However it is possible that our current ice age will change to the normal warm climate Earth that had for most of history.

> "There's been no net global warming in the 21st century. Although seldom reported by the mainstream media, it's quite a story, because no climate model predicted it." – Dr. Marlo Lewis, Jr.

Warming and cooling cycles

During an ice age there are a number of warming and cooling cycles. The cooling periods are called glacial periods and the warming periods are called interglacial periods.

However, to simplify matters, we will just call them cooling and warming periods. These periods have nothing to do with the existence of humans, since they occurred long before humans existed. Humans have been on Earth for only about two hundred thousand years.

During our current ice age there have been regular cooling and warming periods. For the last million years, the cooling periods have lasted about ninety thousand years and the warming periods have lasted an average of about eleven thousand years. We are now in a warming period that started about eleven thousand years ago.

> "Climate change is a natural phenomenon. Climate changes all the time. That fact is not a threat, because, in the past human beings have adapted to all kinds of climate changes." - Dr. S. Fred Singer, atmospheric and space physicist

Our current warming period

Before this current warming period started, most of North America and even my hometown of Chicago were buried in about two miles of ice. Since much of the Earth's water was tied up in the ice, the world's oceans were more than an amazing 400 feet lower than they are today.

When the ocean level was so low, a land bridge formed that joined present-day Alaska and Russia. Humans could now cross the Bering Strait and travel to America. With the oceans now high, that land bridge has disappeared.

We are approaching the end of the warming period and a cooling period could start at any time. Remember that this is all occurring within our current ice age.

> "Nothing that is occurring in weather or in climate research at this time can be shown to be abnormal in the light of our knowledge of climate variations over geologic time." - Thomas Gray, researcher and former head Space Services branch at the NOAA

The following chart shows that Earth has cooling periods that last almost 100,000 years. These cooling periods are punctuated by warming periods that typically last about 11,000 or 12,000 years. Because our current warming period has lasted approximately 11,000 years, the odds are good that a new cooling period will begin in the near future.

Notice that our current warming period is very similar to previous warming periods. And you can see that long before humans existed, the climate was constantly changing. It is obvious that large changes in our climate are definitely normal.

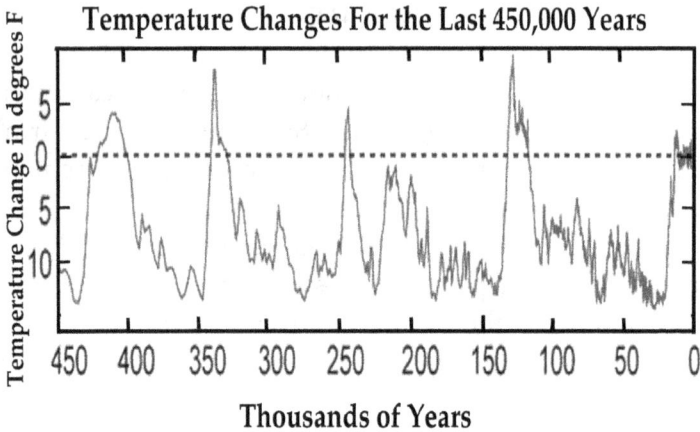

Temperature Changes For the Last 450,000 Years

Thousands of Years

The above chart has the temperature on the left and the thousands of years on the bottom. Starting on the right with zero, which is our present time, you can see that our average temperature is still cooler than it was 120 thousand years ago when we were in the previous warming stage of our current ice age.

The chart does not show actual measured temperatures because for almost the whole period of the chart, thermometers were not invented. Instead the chart comes from computer models of things like ice core samples and tree rings.

The chart indicates that Earth is not yet as hot as it has been in some previous warming cycles. But it also indicates that we could begin another cooling period at any time. The cooling part of the cycle takes much longer than the warming part.

> "You know, to think that we could affect weather all that much is pretty arrogant. Mother Nature is so big, the world is so big, the oceans are so big – I think we're going to die from a lack of fresh water or we're going to die from ocean acidification before we die from global warming." - Chad Myers, meteorologist

Past temperatures are just approximations and can't be called completely accurate. Even in recent times, land-based surface thermometers have not been uniform.

Meteorologist Anthony Watts took a survey of more than 860 temperature stations used by climate scientists. He found them located next to hot roads and asphalt parking lots, near hot exhaust fans of air conditioners, next to very hot rooftops, etc. Some 89% of the stations failed to meet the placement rules for temperature stations. We must conclude that land-based stations can't be called reliable.

> "The Earth has been warming for 12,000 years. If this happened once, and we were the cause of it, that would be cause for concern. But glaciers have been coming and going for billions of years." - Bruno Wiskel, Geologist, University of Alberta

Earth's future temperatures
Earth has been getting colder for the last few years, so the cooling cycle may have already started. 2013 was one of Americas largest one-year temperature drop. The total ice in the Antarctic sea ice is at a recent high, and the Arctic sea ice is near record thickness.

18

> "So far, real measurements give no ground for concern about a catastrophic future warming." - Jarl R. Ahlbeck, scientist, a chemical engineer at Abo Akademi University in Finland, former Greenpeace member

Earth's accurate temperature records only go back to about forty years ago when satellites began taking accurate readings. Compare forty years to Earth's climate history that is measured in millions of year cycles. It is obvious that real temperature readings of Earth are very recent.

Sooner or later, when Earth begins serious cooling, much of Earth will be again covered with ice. Food production will fall, possibly triggering wide spread famine and starvation. It's a fact that food does not grow on top of ice glaciers. But that will take many years. I have no doubt that by then humans will find new ways to grow or synthesize sufficient amounts of food. We don't have to spend trillions of dollars to fight global warming or global cooling. It happens by itself.

> "NASA should be at the forefront in the collection of scientific evidence and debunking the current hysteria over human-caused Global Warming. Unfortunately, it is becoming just another agency caught up in the politics of global warming, or worse, politicized science." - Walter Cunningham, NASA astronaut/physicist, Apollo 7

It is not our fault

Similar cooling and warming cycles have occurred during the entire 2,580,000 years of our current ice age. Since humans have only been on Earth about 200,000 years, it is completely obvious that we are not the cause of global climate change. It is not our fault. We will have to blame Mother Nature.

It is also worth noting that during our current ice age, the Earth's temperature has been cooler than it was during more than 99% of Earth's previous history. So it is completely normal for Earth to be hotter that it is now. And it's completely normal to have global climate change.

No human-caused warming that is distinct from natural system variation and noise has ever been detected.

The "Little Ice Age"

The Medieval Warm Period lasted from about 800 to about 1300 A.D. Temperatures were considerably warmer than they are now. Yes warmer than today.

During that time, the average worldwide temperatures were at least three degrees higher than the average temperatures during the 20th century.

But then in about 1300, a cooling cycle began. It got so cold that many rivers iced over and people had problems keeping warm. This period became known as the "Little Ice Age".

The reason I put the "Little Ice Age" in quotes is because the entire period was within the warming part of our current ice age. But the fact remains that it was much colder during that period then it is today.

During those four centuries of deep bitter cold, glaciers advanced far beyond their previous boundaries. Then during the early 1800's, the ice began to retreat and the world began to warm up. It has been warming ever since.

Alarmists don't like to talk about the end of the "Little Ice Age" because it ended more than 100 years before cars or the industrial age existed in large numbers.

That proves that man-made carbon dioxide emissions had little impact on the current climate change. The warming that started with the end of the "Little Ice Age" destroys the man-made global warming theory.

> "All this argument, is the temperature going up or not, it's absurd. Of course it's going up. It has gone up since the early 1800s, before the Industrial Revolution, because we're coming out of the Little Ice Age, not because we're putting more carbon dioxide into the air." - Dr. Reid Bryson, Meteorologist

During the "Little Ice Age" weather change varied across locations. For example, there were periods when Central and Eastern America had bitterly cold weather. At the same time, Western America was relatively warm. This was due to the effect of ocean currents.

Note the Medieval warming period in the graph on the following page. Climate alarmists have been trying to delete the inconvenient Medieval Warm Period from the Earth's climate history for at least a decade. But they have obviously been unsuccessful.

Below is a graph showing the average temperatures during the last three thousand years of the warming period of our current ice age. Notice that we are no longer in the "Little Ice Age" but all 3,000 years are in the warming period of our current ice age.

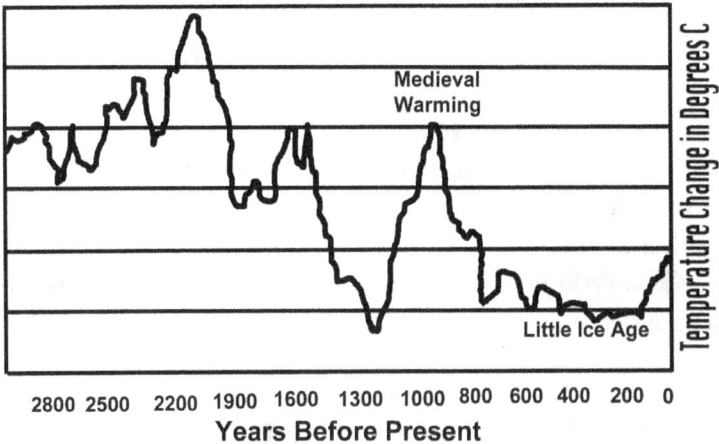

Years Before Present

2800 2500 2200 1900 1600 1300 1000 800 600 400 200 0

"Global warming is largely a natural phenomenon. The world is wasting stupendous amounts of money on trying to fix something that can't be fixed." - Dr. David Bellamy, Botanist and famed environmental campaigner

Chapter Summary

During our current ice age, Earth has had some very warm periods and some very cold periods. It has been warming and cooling long before the existence of humans. Our current average temperature is in line with historical ice age temperatures and is in completely normal ranges.

This one fact alone is enough to prove that our global warming and climate change is not made by humans!

Given a choice between going back to Earth's normal hot temperatures or returning to the cold part of our current ice age, every thinking person would choose normal hot temperatures. During the cold ice age, much of the world would be covered by ice and could not grow food. The result would be mass starvation for many humans.

There are numerous cooling and warming periods during an ice age. We are currently in a warming cycle of Earth's fifth ice age. It has already lasted more than 11,000 years and will likely end within the next thousand years. Earth has recently been getting colder, and we may be heading for a return to a cooling cycle or this may just be another colder period in our warming cycle.

Everything in nature is in constant change. Often this change occurs too slow to notice, but it is always changing. There is no more change today than there has been in the past.

Changes in Earths climate cycles are not your fault. They are just the results of the forces put in play by Mother Nature before Earth even existed, and the changes are beyond human control.

Second Fact

Carbon dioxide is not the number one greenhouse gas.

What is a greenhouse?

A greenhouse is a glass or plastic structure that lets the light and heat from the Sun enter but not leave. The glass or plastic barrier restricts the movement of the warm air inside and does not allow it to mix with the cooler air outside. The efficiency of the heating in a greenhouse is due to the trapping of the air so that it can't escape through convection.

The Earth is not a greenhouse because hot air rises and mixes with colder air. In the open atmosphere, hot air trapping is not applicable. The hot air rises and the heat escapes into space so the term greenhouse does not properly apply.

However, certain components of air are not completely transparent to heat radiation and absorb it on its way upward from the Earth. So a very small percent of the heat is reflected back to Earth. These gases have been given the misleading name of greenhouse gases. These gases are actually much more beneficial for our life on Earth than detrimental. As a mater of fact, without these gases life on Earth would not exist.

What is a greenhouse gas?

The main greenhouse gas is water vapor. It varies from about five percent over warm water to about one tenth of one percent over dry deserts or very cold climates. The average is about two percent. It is constantly changing, however, so there is no real accurate number. Water covers over seventy percent of the Earth and that provides for a lot of evaporation.

"The issue of the 'greenhouse effect' has assumed a peculiar life of its own... At the same time, one has no difficulty hearing the muttering in the corridors of any meteorology department that this is an issue that has gotten out of hand, that the claims are insupportable, that the models are inadequate, and the data contradictory." - Alfred Sloan, Professor of Meteorology at the Massachusetts Institute of Technology,

Water vapor accounts for about ninety-five percent of the "greenhouse" effect. The other greenhouse gases are just trace gasses and together account for only about five percent of the greenhouse effect.

There are a number of other minuscule components of the atmosphere such as carbon dioxide, but they are all very small trace gases and relatively insignificant compared to water vapor.

"There is no observational evidence that the addition of human greenhouse gas emissions have caused any temperature perturbations in the atmosphere." - Dr. George T. Wolff, former member of the EPA's Science Advisory Board, and authored more than 90 peer-reviewed studies.

There are also a lot of facts that most people don't know about water vapor. For example, the amount of water vapor near the North and South poles is very low but the amount of water vapor near the equator is about five percent. And that is where the Sun's rays are strongest. So water vapor being concentrated near the equator accounts for a much larger part of the "greenhouse effect."

The Sun is weakest at the poles where there is little water vapor. But even if there was more water vapor, there is so little Sun at the poles that it would hardly matter.

The alarmists ignore this information because they would not get the results they want. Their computer models do not show water vapor, even though it is by far the absolute number one most important climate change gas in the world. Alarmists use bad science.

> "'Scientific' computer simulations predict global warming based on increased greenhouse gas emissions over time. However, without water's contribution taken into account they omit the largest greenhouse gas from their equations. How can such egregious calculation errors be so blatantly ignored? This is why man-made global warming is 'junk' science." - Michael Myers, analytical chemist, specializes in spectroscopy and atmospheric sensing

In my computer classes we had an expression for this bad information: "garbage in makes garbage out." It is unfortunately true that the alarmists base their entire theory on untested theoretical computer models of the climate. And these models are just plain wrong.

> "Since I am no longer affiliated with any organization nor receiving any funding, I can speak quite frankly...As a scientist I remain skeptical." - Joanne Simpson, former elite NASA climate scientist

Chapter Summary

Carbon dioxide's greenhouse effect is minuscule compared to water vapor, which is the big elephant in the room. The main greenhouse gas is water vapor. Water vapor accounts for about ninety-five percent of the "greenhouse" effect.

This information is entirely skipped by the alarmists because they would not get the results they want. Their computer models do not show water vapor, even though it is by far the absolute number one most important climate change gas in the world. In effect, their computer models are useless.

Third Fact

Carbon dioxide is our friend.

Carbon dioxide is our friend. Without it life on Earth would not be possible. Earth and all its life forms need more carbon dioxide and not less.

"Carbon dioxide is absolutely vital for living on Earth. It's plant food, and all of life lives off carbon dioxide. To demonize it shows that you don't understand school child science." - Ian Plimer, geologist, professor emeritus of Earth sciences

Carbon dioxide is not a poison or a bad gas. As humans we all make it in our bodies and release it every time we exhale. Politicians hope that we will confuse it with carbon monoxide, but they are not the same at all.

Unlike carbon dioxide, carbon monoxide is a component of auto exhaust and can kill humans by combining with our red blood cells and making them unable to carry oxygen throughout our body. However carbon dioxide is a completely different chemical from carbon monoxide.

Carbon monoxide is the result of incomplete combustion and is a pollutant. Fortunately, carbon monoxide breaks down very quickly in the atmosphere. It only lasts about a month or two. It rapidly combines with oxygen to form harmless carbon dioxide. So it is only in enclosed areas where it could be a problem.

In contrast, carbon dioxide is not a pollutant or smog. (Pollutant components include sulfur dioxide, nitrogen oxides, particulate matter, ozone, and carbon monoxide, but not carbon dioxide.) Possible future increases in carbon dioxide will not produce any adverse health effects.

Carbon dioxide is green. More carbon dioxide in the air means more plant growth. Less carbon dioxide is brown and means less plant growth.

> "Carbon dioxide is an insignificant component of the earth's atmosphere and that, rather than being the purveyor of doom it is currently viewed as today, it is needed in order for plants to grow." - Dennis Hollars, astrophysicist

We use carbon dioxide in many ways, and it is even added to our soft drinks to make them bubble. If it were a poison we wouldn't add it to our drinks. Even dry ice, used to cool our foods, is almost pure carbon dioxide.

Carbon dioxide, sunlight and water are the main ingredients required for photosynthesis and plant growth.

Carbon dioxide is a plant food and is necessary for green plant growth. Some politicians say we need to reduce carbon dioxide by 50% to save the earth. In fact if levels decrease below 50% of today's levels, photosynthesis will drastically slow down. Then many green plants will turn brown and die causing widespread famine and death. Even at 70% of current carbon dioxide levels, a slow-down in plant growth is significantly noticeable. So more carbon dioxide is green and less carbon dioxide is brown.

> "I am convinced that the current alarm over carbon dioxide is mistaken...Fears about man-made global warming are unwarranted and are not based on good science." - Dr. Will Happer, Professor of Physics at Princeton University

Earth would do better with more carbon dioxide and not less. Historical increases in carbon dioxide have resulted in increased food supplies. Plants actually grow faster, bigger, and better with more and not less carbon dioxide in the air. Over 5,000 peer reviewed studies show the amazing increase of growth in plants that are grown in high carbon dioxide environments.

By adding about 60% more carbon dioxide to the air inside a real greenhouse, growers have been able to get a 40% crop increase in food growth and plant size. Some studies show that the optimal level for many green plants is about 300% greater than found in normal air.

> "Carbon dioxides ability to trap heat declines rapidly, logarithmically, and reaches a point of significantly reduced future effect explaining why correlations with carbon dioxide don't hold. A far more consistent and significant correlation exists between the planet's temperature and the output of energy from the Sun." - Leighton Steward, geologist, twice chaired the Audubon Nature Institute

Carbon dioxide does not stay in the atmosphere forever. Mother Earth is constantly scrubbing carbon dioxide out of the atmosphere. Green plants remove carbon dioxide and use it to grow. So the more carbon dioxide in the atmosphere, the more is removed by plant life growing faster and bigger.

A study by the National Oceanic and Atmospheric Administration conducted between 1988 and 1992 showed that America's forests were very effective at regulating carbon dioxide in the atmosphere. They reported that the amount of carbon dioxide absorbed by the trees was much more than all the emissions from America's cars, power plants, and factories.

> In fact, there is only 1/19 as much carbon dioxide in the air today as there was 520 million years ago." - Greg Benson, Earth scientist, geologic study/geologic modeling

Another major way Mother Nature regulates carbon dioxide in the atmosphere is through weathering. Carbon dioxide reacts with limestone (calcium carbonate) to form water-soluble calcium bicarbonate. This soluble compound is then washed away with the rainwater. This form of weathering substantially reduces the amount of carbon dioxide in the air. Again, the more carbon dioxide in the air the more weathering occurs.

It is estimated that the residence time of carbon dioxide in the atmosphere is about seven years. So carbon dioxide does not last forever and is constantly being removed.

> "Man-made emissions of carbon dioxide are clearly not the source of atmospheric carbon dioxide levels." - Dr. Murry Salby, Professor and Climate Chair at Macquarie University, Australia

So the more carbon dioxide in the air, the faster and bigger the trees will grow and the more weathering will occur. This helps to balance the total carbon dioxide levels.

Carbon dioxide is a harmless, colorless, and odorless gas. It is naturally occurring and human life would not exist without carbon dioxide. Carbon dioxide is also the gas that makes our bread rise.

> "For most of Earth's history the carbon dioxide level has been several times higher than the present." - Dan Pangburn, mechanical engineer, author of a climate research paper

Carbon dioxide is a fertilizer, and as I indicated some real greenhouses add it to their air to increase plant growth. We absolutely need carbon dioxide in the air. It is not a poison. It is an essential gas.

With the stronger plant growth of higher carbon dioxide levels also comes a better resistance to various plant stresses. It is as if the plants are healthier and do better in almost every area.

Surprisingly, higher concentrations of carbon dioxide create plants that are healthier to eat. Certain health-promoting substances, such as vitamin C and antioxidants, are increased with higher levels of carbon dioxide.

Just as grains are called the staff of life for humans, carbon dioxide is the staff of life for our green plant kingdom. Today however, because schools propagate political science, children are more likely to think of carbon dioxide as a poison to be avoided.

Chapter Summary

Carbon dioxide is not a poison or a bad gas. Carbon dioxide is not a pollutant or smog. Carbon dioxide is a plant food and is necessary for green plant growth and for our very lives. Earth would do better with more carbon dioxide and not less.

Fourth Fact

Almost all carbon dioxide comes from natural sources and not human activity.

Parts Per Million

Would you rather have four hundred parts per million of a million dollars, or four hundredths of one percent of a million dollars?

Almost every one I asked would rather have the four hundred parts per million. They don't really know why. They think it is to difficult to actually figure out, but the words hundred and million sound like much more money then the word hundredths. So they pick million.

In truth, both are exactly equal. The reason this concept is important is that climate alarmists use "parts per million" when talking about carbon dioxide. To the non-scientist four hundred parts per million sounds like a lot, but in truth carbon dioxide is just a very small trace gas.

In the chart below you can see that carbon dioxide is a very small component of our air. These percentages change slightly depending on the temperature area and the time of the year.

Water vapor varies by volume in the atmosphere from a trace to about 4%. Therefore, on average, only about 2 to 3% of the molecules in the air are water vapor molecules. - Jeff Haby, Meteorologist

Nitrogen – 76.51%
Oxygen – 20.52%
Water vapor – 2.00%
Argon – 0.92%
Trace gas including: Carbon Dioxide, Neon, Methane, Helium, Krypton, Hydrogen, etc. Total trace gas – 0.05%

Carbon dioxide is a trace gas. It is a very minor component of the atmosphere, less than four hundredth (0.04) of one percent. Compare that with water vapor that averages about two percent of the atmosphere. This means there is 5,000 percent more water vapor in the air than carbon dioxide. And Mother Nature and not humans cause almost all the carbon dioxide.

Compared to water vapor, the "greenhouse" effects of carbon dioxide are negligible. It is too meager a part of our atmosphere to have any real significance.

When sunlight strikes Earth, its energy is converted mostly to heat. And some of that heat is reflected back to outer space. The atmosphere reflects some of this heat. Eventually, all the energy is radiated back into space.

The band of light that a carbon dioxide molecule can absorb is much smaller than the band of light a water vapor molecule can absorb. This means a given quantity of water vapor absorbs twice as much energy as the same quantity of carbon dioxide. However, as we just learned, they are not in equal quantities. There is tremendously more water vapor than carbon dioxide in the atmosphere.

"Even doubling or tripling the amount of carbon dioxide will virtually have little impact, as water vapor and water condensed as particles in clouds dominate the worldwide scene and always will." – Dr. Geoffrey Duffy, Chemical and Materials Engineering professor

Human contribution

The human contribution to carbon dioxide is only 5% of that 400 parts per million. So the human contribution is only about 20 parts per million. This is so small it's not even enough for a serious discussion of it causing global warming. The alarmist's entire viewpoint is based on very bad math.

Certain people continue to preach that carbon dioxide is the most common greenhouse gas. I have stopped trying to explain reality to these brainwashed souls. Their attitude seems to be: "don't confuse me with the facts."

Unfortunately, existing computer climate models are extremely crude. There's a huge list of natural phenomena that are absent from the models. Just two of these missing phenomena are water vapor and clouds. Also if a model does not get the correct politically acceptable results there will be difficulty getting funded and the model will quickly change or disappear.

"Carbon dioxide is not the major greenhouse gas. The major greenhouse gas is water vapor. But current climate models do not know how to handle water vapor and various types of clouds. That is the elephant in the corner of this room." - Lowell Wood, astrophysicist

Natural contribution

Almost all carbon dioxide comes from natural sources that occurred before humans existed. Humans account for only five percent of the carbon dioxide in the atmosphere. Mother Nature is responsible for the other 95 percent.

Mother Nature's carbon dioxide comes from many sources, such as volcanic eruptions, forest fires, and normal plant decay.

"The solubility of CO_2 in water decreases as water warms, and increases as the water cools. The warming of the earth since the Little Ice Age has thus caused the oceans to emit CO_2 into the atmosphere." – Dr. Howard Hayden, retired professor of physics at the University of Connecticut

However, the main source of change in the amount of carbon dioxide is the ocean. The ocean is huge and has enormous amounts of carbon dioxide dissolved in the water. When the ocean gets warmer, the carbon dioxide dissolved in the ocean evaporates into the atmosphere.

This is much the same as the carbon dioxide that is dissolved in carbonated soft drinks. If the soft drink gets warm, it releases its bubbles more rapidly and your drink gets flat. If the soft drink is kept cold, it retains the carbon dioxide longer.

So here we have a situation where carbon dioxide has only a very negligible effect compared to water vapor on the "greenhouse" effect. And humans are responsible for only a very negligible five percent of that total.

Humans have only a very negligible effect on a very negligible gas. So as far as "greenhouse" gas goes, the human effect is very very negligible.

Warming Causes Carbon Dioxide

You read that right. Global warming causes increased carbon dioxide and not the other way around.

Some years ago, the climate alarmists claimed that when the carbon dioxide in the atmosphere increased it was followed by an increase in the heat of our atmosphere. They claimed that ice core samples show a strong correlation with warming temperatures. They concluded that when carbon dioxide goes up then the temperature gets hotter.

This mistake in logic is similar to the belief that when the rooster crows, then the Sun rises. Therefore, the rooster causes the Sun to rise.

We all know correlation does not mean causation. A more carful reading of the those ice core samples shows that after the Earth warms up first, then more carbon dioxide is released into the atmosphere. It is similar to a soft drink that loses more carbon dioxide bubbles when it is warm than when it is cold.

In addition there is an enormous amount of hydrocarbons such as leaves on the ground. When it is warmer they decay faster and decay is just a very slow burning. The result is heat creates carbon dioxide. When these leaves are covered by ice or snow they decay much slower.

So the increased heat in the atmosphere causes increased carbon dioxide and not the other way around. The alarmists don't talk much about ice core samples any more.

> "So far, real measurements give no grounds for concern about a catastrophic future warming." - Dr. Jarl R. Ahlbeck, a chemical engineer and former Greenpeace member.

Over the past hundreds of thousands of years, the carbon dioxide increase lags global warming by about a hundred years. This means that warming temperatures come first and cause increased carbon dioxide. Carbon dioxide does not cause global warming. During Earth's past it was not at all uncommon to see carbon dioxide levels much more than double today's readings.

The theory of the climate change alarmist argument was that rising carbon dioxide causes global warming. But that theory was never proven and it is neither viable nor true.

> "We thus find ourselves in the situation that the entire theory of man-made global warming (with its repercussions in science, and its important consequences for politics and the global economy) is based on ice core studies that provided a false picture of the atmospheric carbon dioxide levels" - Dr. Zbigniew Jaworowski, Physicist

From million year old ice-core records taken in Greenland and Antarctica, scientists can observe the approximate temperature and carbon dioxide levels in the air. The ice-core samples showed that there was a correlation between the carbon dioxide level and the global temperature.

At first, the samples were thought to show that carbon dioxide causes global warming. As I previously indicated, however, they actually showed that the warming came first and caused the increased carbon dioxide.

"The Kyoto theorists have put the cart before the horse. It is global warming that triggers higher levels of carbon dioxide in the atmosphere, not the other way around." - Andrei Kapitsa, Antarctic ice core researcher.

Even much shorter time period comparisons of temperature and carbon dioxide shows that the temperature change comes first and the carbon dioxide follows the temperature. The highlights of the following recent study by three prominent scientists shows that shorter term temperature changes also cause carbon dioxide changes.

"The maximum positive correlation between carbon dioxide and temperature found carbon dioxide lagging 11–12 months in relation to global sea surface temperature, 9.5-10 months to global surface air temperature, and about 9 months to global lower troposphere temperature." - Ole Humluma, Department of Geosciences; Kjell Stordahlc, Department of Geology; Jan-Erik Solheimd, Department of Physics and Technology

What makes Carbon Dioxide?

There are many sources of carbon dioxide. However the largest source is the oceans. They are the great storehouse of carbon dioxide. They have a huge effect on its concentration. The ocean alone can significantly raise or lower the carbon dioxide in the air.

Hundreds of ice core samples show that the climate warming leads the carbon dioxide increases by about a hundred years or more. This means that it is the warmer temperature changes that cause increased carbon dioxide. The climate alarmists' entire case rests on the belief that increased carbon dioxide causes increased climate warming. But they are wrong and it is the other way around.

The oceans were getting warmer and that put more carbon dioxide into the air. However, recent measurements have shown that the oceans are no longer getting warmer.

"Atmospheric carbon dioxide is a dynamic stream, from the warm ocean and back into the cool ocean. Public policy represented by the Kyoto Accords and the efforts to reduce carbon dioxide emissions should be scrapped as wasteful, unjustified, and futile." - Jeffrey A. Glassman, physicist and engineer

Most of the climate alarmists just want to do the right thing and really believe that carbon dioxide causes global climate change. But it doesn't. And now we are getting more and more of those ugly windmills destroying our view of the countryside. And did you forget about all those birds that are getting killed by the whirling blades?

Cooling and warming periods have been coming and going for the entire existence of Earth. Billions of dollars of government grants and research money are being funneled into promoting climate alarmism. If you spend money on things that can't be changed, the money is wasted.

> "The cleverest thing that the global warming alarmists have done is to categorize carbon dioxide emissions as pollution, because it's not true." - Myron Ebell, Director of International Environmental Policy at the Competitive Enterprise Institute

Chapter Summary

Changes in atmospheric carbon dioxide follow changes in Earth's temperatures. As our oceans heat up, more carbon dioxide is released into the atmosphere. A cause does not follow an effect. So the cause of changing carbon dioxide is Earth's changing temperature.

It's completely illogical to believe in an apocalypse caused by human carbon dioxide contributions.

Fifth Fact

The Sun causes climate change

So what causes global climate change? It should be obvious that virtually all of the heat on Earth comes from our Sun. So it is our Sun that supplies Earth's heat. Any change in the Sun's output would certainly affect Earth's average temperature.

> "Global warming - at least the modern nightmare vision - is a myth. I am sure of it and so are a growing number of scientists. But what is really worrying is that the world's politicians and policy makers are not." - David Bellamy, biologist

Because our lives are so short compared to geological time, it is difficult for humans to realize that our Sun is constantly changing its heat output. The Sun's output is not a constant but a variable factor. It has had changing energy output long before the Industrial Age and long before humans existed.

The Sun's energy changes in many ways. Among them are magnetic fields, cosmic rays, brightness, and eruptive activities. Some of these variables occur in cycles of 11, 88, 106, 213, and many more years.

The Sun has warmed everything on Earth since life began. We can safely say that life as we know it would not exist on Earth if it were not for the heat of the Sun. And for thousands of years humans have know that their lives

44

depend on the Sun. In fact many cultures have worshiped it and even have Sun Gods.

Everywhere in our universe we observe cyclical variations. And our Sun and Earth are no different. For example, Mars seems to get warmer when Earth gets warmer.

Cyclical changes in solar activity are obviously nothing new. The cycles of our Sun are a major cause our climate change and our ice ages.

> "Sunspot cycles and their effects on oceans correlate with climate changes. Studying these and other factors suggests that a cold, not warm, climate may be in our future...Some scientists believe that an extreme cooling episode, potentially a mini-ice age, is imminent. Others think that it may already be under way." – Dr.Joseph D'Aleo was a professor of Meteorology and the first Director of Meteorology at the cable TV Weather Channel

It's a fact that the Sun is going through many cycles of different durations. Some, like sunspots, are short eleven-year cycles. Many of the cycles that cause global heating and cooling are very much longer. There are also changes in the types of radiation, solar winds, and particles that the Sun produces.

> "The cause of these global changes is fundamentally due to the Sun and its effect on the Earth as it moves about in its orbit. Not from man-made activities." - Dr. William W. Vaughan, Retired Award Winning NASA Atmospheric Scientist

The Sun is very complicated and humans still do not fully understand it. We know that its core occupies only 2 percent of its total volume, but in it is concentrated about half the total mass of the Sun. The core is where the thermonuclear fusion of the Sun takes place. However, this energy must make its way up to the Sun's surface and, depending on the charged plasma and ever-changing huge magnetic fields, this could take many thousands of years. But once the energy gets to the surface it takes only eight minutes to get to Earth.

History has shown us that when sunspot activity increased, the earth got warmer. And when the sunspot activity decreased, the earth got colder. Decreased sunspot activity indicates less magnetic activity in the Sun. This means that the Sun's surface is not releasing extra heat.

Our last eleven-year sunspot cycle has not restarted on time, which means less heat is coming from the Sun. This would typically indicate future lower global temperatures and possibly the beginning of a new cooling period.

The Sun is measurably dimming and its output has dropped in comparison with the levels at the last solar minimum. Its magnetic field is also at less than half strength. Relatively small changes in solar activity can affect Earth's climate in complex but significant ways.

If the current trend of the dimming of the Sun continues, it is entirely possible that it will lead to the long-expected transition back to the cooling part of our current ice age. When exactly the transition to the cooling cycle will happen is still unknown.

But judging by the Sun's decreased energy, we will experience weather slightly cooler than we have had in the first few years of this century. Not enough is known about the Sun, so it is possible we will suddenly enter a period of increased solar activity. Only time will tell if our cooling cycle has started or if this is a just temporary change in solar heat output.

> "The world is wasting stupendous amounts of money on trying to fix something that can't be fixed. The climate-change people have no proof for their claims. They have computer models which do not prove anything," - David Bellamy, conservationist and TV personality

Except for very recent history, there was no actual documentation available that tells us the amount of energy at various wavelengths coming to Earth from our Sun. Consequently, we must use incomplete computer models, data from polar ice cores, and other approximations to understand solar energy output.

With the beginning of our warming cycle over eleven thousand years ago, civilization also began to flourish. Eleven thousand years ago, the world's population was about three million people. Because of the warming cycle, the population surged and today it is over seven billion people.

The Sun is huge and weighs more than 300,000 times the weight of Earth. It is basically an enormous hydrogen bomb that is constantly exploding, held together by its own tremendous gravity, pressure, and magnetic forces. When the temperature emitted by the Sun changes, our temperature here on Earth also change.

"This treaty [Kyoto] is, in our opinion, based upon flawed ideas. Research data on climate change do not show that human use of hydrocarbons is harmful. To the contrary, there is good evidence that increased atmospheric carbon dioxide is environmentally helpful." - Frederick Seitz, Past President, U.S. National Academy of Sciences

It has just been within my lifetime that science has developed the instrumentation to study the various radiation frequencies emitted by the Sun. So our scientific observed history of our Sun is quite recent.

Changes in different radiation frequencies can also cause temperature changes. Changes in the Sun's ultra violet light affects cloud formations. Clouds make a substantial difference in the heat that gets through to Earth.

Our Sun provides the heat and light that make life on Earth possible, yet Earth receives less than a billionth of the energy that leaves the Sun. So even slight changes in the Sun could cause Earth to get much hotter or colder.

For many years, scientists have established and studied the relationship between various solar activities and climate change cycles. However, the United Nations Intergovernmental Panel on Climate Change as well as most of the mass media pretends that this relationship does not exist. Obviously it would detract from their carbon dioxide crusade. But our current downward trend in solar activity will soon force the alarmists to invent additional excuses for the lack of "disastrous global warming."

Sunspots

Sunspots are the many dark areas of irregular shape that appear on the Sun. They are caused by incredibly powerful and complicated ever-changing magnetic fields. Sunspots exist in pairs that have opposite magnetic fields.

The result of sunspots is that the heat the Earth receives from the Sun varies. Sometimes there are many years of prolonged sunspot minimums. From about 1650 to about 1720 there were very few sunspots. This coincides with a period of global cooling that records say occurred about the same time.

This was a difficult period for European civilization. It is estimated that famine took the lives of over ten percent of the population. Violent storms and massive flooding caused many additional deaths.

Today, the fast pace with which solar activity is declining suggests that solar cycles without sunspots are on the way. If the pattern holds, we may soon see a period of progressively colder temperatures comparable with the temperatures of the "Little Ice Age."

Obviously there are even longer but inadequately understood periods of diminished or increased heat that cause even more substantial changes in global temperatures.

On the next page is a graph showing the average temperatures during the last ten thousand years of the warming period of our current ice age. Notice that we are no longer in the "Little Ice Age," but all 10,000 years are in the warming period of our current ice age. And most of the time the average temperature was warmer than it is currently.

Temperature Last 10,000 Years

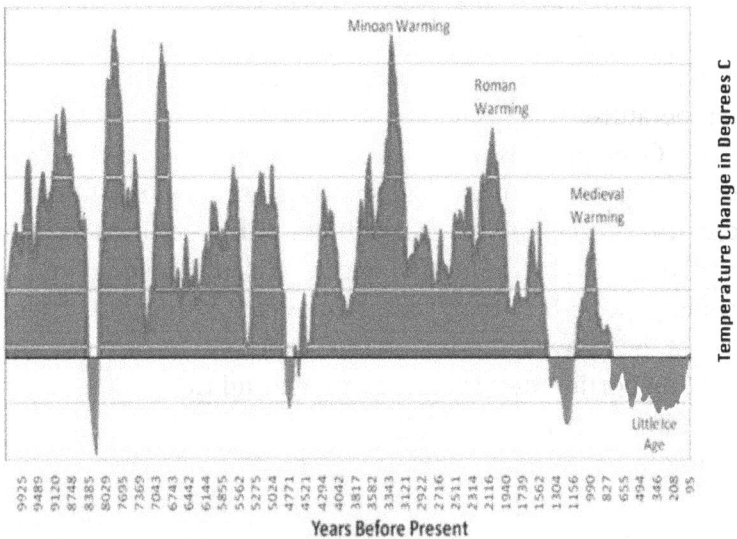

Years Before Present

Notice that in this ten thousand year chart the temperatures in the last three thousand years have been trending lower. Recent uncommon solar activity indicates that the Sun will be less active for years to come. This includes fewer sunspots, missing solar jet streams, and weakening magnetic activity near the Sun's poles. We are now in a longtime downward trend in solar activity. It is possible that this will cause a corresponding climate cooling trend for at least twenty years.

Less energy radiated to Earth result in colder temperatures. There are conflicting theories as to how long these colder temperatures will last. They could end in eleven years or they could herald the beginning of another cooling cycle in our current ice age. If that is the case, the temperatures could last as long as ninety thousand years. My bet, however, is that they will last about thirty years.

Chapter Summary

Almost all of the heat on Earth comes from our Sun. So it is our Sun that supplies Earth's heat. Any change in the Sun's output would certainly affect Earth's average temperature.

Over periods of thousands of years, the Sun itself is changing. At this point in time the Sun is getting dimmer and less energetic. This means that Earth is getting less heat and that affects the entire Earth. We are moving into a cooling period. It may only last twenty years. But it also might be the beginning of our expected move into the cooling portion of our current ice age and the cooling could last ninety thousand years.

Sixth Fact

Volcanoes, asteroids and meteorites can cause climate change.

Volcanic eruptions have had significant effects on Earth temperatures. In 1783, Benjamin Franklin said that the volcanic dust coming from Iceland was responsible for the unusually cool summer. In addition to dust, the volcanic eruption also released large quantities of sulfur dioxide, which does more to block sunlight than actual dust. There have been many similar occurrences.

Volcanic eruptions have been said to cause significant cooling on Earth. An eruption in 1883 created volcanic winter-like conditions. Because of the eruption, the following four years were unusually cold.

Volcanic eruptions also release huge quantities of carbon dioxide. Sometimes they contributed to carbon dioxide levels that were three times greater than they are today.

But the cooling results of volcanic eruptions have not been known to last for centuries. Therefore, all else being equal, they do not cause ice ages. However, volcanic eruptions do cause short-term climate changes and contribute to natural climate variability.

Meteorites

Meteors strike the Earth almost every day. However, meteors of less than 15 feet in diameter usually disintegrate before actually reaching the surface. Meteorites are bigger than meteors and do not completely burn up in the atmosphere. Parts of the meteorites survive to hit the ground.

A large meteorite or asteroid could collide with the Earth, producing the equivalent force of millions of the nuclear bombs. They can form huge craters as they collide with the ground.

The combination of the debris from the meteorite or asteroid and the huge chunks and large quantities of dust of Earth is forcefully ejected high into the atmosphere. Vaporized material can form an incredible aerosol cloud. The intense heat can cause multiple widespread firestorms. This puts even more smoke and debris into the atmosphere. This can easily block light from the sun for long periods of time.

Earth would immediately start to cool. Depending on the size of the asteroid, Earth could experience an "impact winter." The lack of sunshine can cause many green plants to die. This would result in diminished food supplies and soon animals would begin to die.

However, the air would clear up after a few years and sunlight would begin to warm the Earth. So by themselves, meteorites or asteroids do not cause ice ages to start.

What killed the dinosaurs?

Dinosaurs first appeared about 231 million years ago, and were the dominant vertebrates for many million of years. During their existence, there were no ice ages and no polar ice caps. The oceans were estimated to have been even higher than they are today. Atmospheric temperatures were also much higher then they are today.

About 66 million years ago, all of the dinosaurs suddenly became extinct. It is believed that a huge asteroid collided with Earth, creating an "impact winter." Since humans first appeared only about 200 thousand years ago, humans never came into contact with the dinosaurs.

There are a few different theories of exactly what happened to the dinosaurs. But all the theories believe that Earth's temperature became very cold. The majority opinion is that a very large asteroid about six miles in diameter slammed into the earth, releasing an explosion approaching the energy of a billion nuclear bombs. Many of the dinosaurs would have died immediately. Other dinosaurs could have lived for a while, but the atmosphere quickly became much colder.

It is thought that the asteroid strike pulverized the impact area and produced huge quantities of aerosol size particles. These particles were thrown high into the atmosphere and shielded the earth's surface from sunlight. This lack of Sun very quickly decreased Earth's temperature.

With a sudden cold snap and without normal sunlight, most of the plants soon died. Dinosaurs were either cold blooded or had a limited ability to control their body heat. In either case, Earth's cold temperature could have killed them. Dinosaurs also needed large quantities of plant food. The cold and less sunlight would have diminished plant growth. Between the cold and starvation, the remaining dinosaurs did not last very long. Global cooling was the cause of their extinction. Global cooling is much more dangerous than global warming.

This was not the beginning of an ice age. Within a few dozen years, the aerosol size particles diminish and Earth returned to its normal warm weather. Many plants began to grow again, but the dinosaurs and many other species were extinct forever. Earth continued in its normal warm state for the next 64 million years until our present ice age started.

Asteroids similar to the one causing the extinction of the dinosaurs would not cause the extinction of humans. Many, if not most, humans would certainly die. Our food storage and ability to quickly build real greenhouses with artificial light would assure that many humans would survive such a disaster.

> "We theorize that the meteorite strike produced huge quantities of sulfate particles, such as are often blown high into the atmosphere during a volcanic eruption, and these particles shielded the Earth's surface from sunlight. The decrease in solar energy ultimately caused a long cold spell, called an 'impact winter,' that persisted for years." – Dr. Matthew Huber, Purdue University

Chapter Summary

Volcanic eruptions can cause significant damage and temporally cool off the earth or parts of it. However, they have never been the lone cause of an ice age. They may have contributed to the timing of the inception or the end of ice age periods.

Asteroids can also cause significant damage and end the existence of some species. However, they have never caused an ice age to begin.

Seventh Fact

Cosmic rays, micrometeorites and cloud variation can cause climate change.

On balance, clouds produce a cooling effect. As everyone knows from personal experience, when it is cloudy, it gets a lot colder. Without clouds Earth would be very much hotter. The white tops of clouds reflect a significant portion of the incoming solar energy back into space. This causes a substantial energy loss and the consequent is the cooling of Earth.

Clouds are responsible for most of the reflected sunlight from the Earth. Overall, the Earth reflects about 30 percent of the Sun's energy. A small change in cloud cover could make a big difference in the temperature of Earth.

Cosmic rays

Cosmic rays are very high-energy radiation, composed of subatomic particles such as protons and atomic nuclei. They are not actual rays but particles from ancient stars that have exploded. They come mainly from outside our solar system but some also come from our Sun. As our Sun and Earth travel around the center of our galaxy we go through areas of greater or lesser cosmic rays.

Cosmic rays have only been accurately measured for about seventy years. However they appear to have increased since measurements were first recorded. Additionally, it has been experimentally proven that cloud formation is affected by cosmic rays. More cosmic rays means more cloud formation. This coincides with the real world observation of an increase in Earth's cloud formations.

Much less accurate records of cosmic rays and temperature go back 600 million years. They also show a strong correlation between increased cosmic rays and colder Earth temperatures.

Clouds are not water vapor but instead are made up of water droplets. Water droplets only form when there are particles to condense around. These particles could be soot, various kinds of dust particles, a living bacteria, called pseudomonas syringae, micrometeorites or cosmic ray particles.

In addition to making clouds that reduce the sunlight reaching earth, cosmic rays decrease water vapor since it is turned into clouds. And because of its abundance, water vapor is the only significant green house gas. A decrease in water vapor and a decrease in sunlight because of clouds is a powerful cooling combination.

Cosmic rays have been severely underrated in their ability to cause climate change. More research needs to be preformed before we can predict the magnitude of their effect on clouds and climate change.

Micrometeorites

Micrometeorites are extraterrestrial particles that fall to Earth. They are very small, often just a few atoms. Recent measurements show that billions of them reach Earth's atmosphere every single day. But the amounts vary depending on the area of space Earth is in.

Micrometeorites, like cosmic rays are a source of seeds that can turn water vapor into water droplets that form clouds. Micrometeorites are found in rain and that helps us estimate their abundance.

Our Galaxy

Some astronomers believe that there is a connection between Earth's global climate change and our solar systems passage through space and around the Milky Way galaxy.

We know that Earth periodically travels through some of the spiral arms of our galaxy. This takes millions of years but the age of Earth is measured in billions of years. We also know that these spiral arms are associated with higher energy regions in our galaxy.

The entire Milky Way galaxy is circling around its massive center. So Earth is not just moving around the Sun. It is also moving around the galaxy. And additionally the entire Milky Way galaxy is speeding through our universe.

In all that movement, our solar system encounters different energy regions. And these different energies can actually affect the temperatures here on Earth.

One type of energy that bombards Earth is cosmic rays (composed primarily of high-energy protons and atomic nuclei particles). Scientists have found that the strength of cosmic rays has varied through the years. They say this is possibly the main cause of climate change.

Cosmic rays (particles) are thought to be the remains of stars that have exploded in the past. It is logical that as earth travels though space there are sometimes more and sometimes less cosmic rays raining down on Earth.

Cosmic rays can be the seed for the clouds that help cool the planet. As cosmic rays increase they initiate the formation of more clouds. And the more clouds in the sky, the less sunlight gets through to warm Earth. Clouds have a very powerful effect on our climate.

Micrometeorite density also various depending on where Earth is in the galaxy. A significant change in their numbers would definitely make a difference in our weather.

"What the 'scientific consensus' has failed to account for is that global warming or cooling can happen through natural cloud changes altering the amount of sunlight being absorbed by the Earth." – Roy W. Spencer, former senior NASA climatologist

There is also the question of what effect some of these energy regions have on the energy output of our Sun. With all the dark energy and dark matter, the possibilities of climate change in our travels through unknown space are very real. Additionally, a change in the energy transmission properties of space between the Sun and the Earth is also a very real possibility.

"It's just that the press only promotes the global warming alarmists and ignores or minimizes those of us who are skeptical. To many of us, there is no convincing evidence that carbon dioxide produced by humans has any influence on the Earth's climate." – Mark L. Campbell, professor of chemistry

As with water vapor, this information is entirely skipped by the global warming alarmists because they would not get the results they want.

Eighth Fact

Changes in Earths orbit and in our Solar System can cause climate change.

Earth's tilt towards the Sun and its orbit around the Sun is not constant. Our orbit changes both in distance from the Sun and also in shape, moving from a circle to an ellipse and then back again. This changes the average distance between Earth and the Sun. And of course this also changes the average heat from the Sun. These cycles repeat about every hundred thousand years and could easily cause the hundred thousand year warming and cooling periods we see in our current ice age.

Most people do not realize that the other planets in our solar system affect the orbit of Earth, but they do. Even our small moon affects our ocean tides on a regular basis.

The biggest planet in our solar system is Jupiter. It weighs more than twice as much as all the other planets combined. Its weight is about 318 times greater that the weight of Earth. Earth is the third planet from the Sun and Jupiter is the fifth, so they are relatively close together. Jupiter is so large that sixty-seven moons orbit it. Jupiter takes almost ten years to go around the Sun.

Depending on the alignment of Earth with the other planets, our orbit changes, so the distance from Earth to our Sun changes. Because of the laws of physics, a small change in the distance from the Sun makes a much larger change in the total heat received. These changes in distance therefore change Earth's climate.

Earth's changes in orbit are possibly not strong enough by themselves to trigger a new cooling phase in our current ice age. However, orbit changes in combination with predicted solar changes or a large asteroid strike could easily put us over the edge into a long cooling period.

> "Solar radiation was the trigger that started the ice melting, that's now pretty certain." – Peter Clark, professor, Oregon State University

Some climate researchers believe that the wobbles in Earth's rotation cause the cooling and warming cycles in our current ice age. They believe the periodic temperature changes were caused by changes in solar radiation and have nothing to do with changes in carbon dioxide levels.

Our Solar System

Earth does not travel around the Sun in a perfect circle. It travels in an ellipse so sometimes it gets closer or further from the Sun. Earth's farthest distance from the Sun is about 94.4484 million miles. Its closest distance is about 91.3416 million miles. The eccentricity and total distance are not in fact constant and vary on timescales of about 100,000 years because of the influence of other planets (mostly Jupiter) on Earth's orbit.

If you calculate time in place with orbit distance you will find that there is a variation in average energy from the Sun ever 100,000 years.

So once about every 100,000 years our orbit gets millions of miles further from the Sun, and once every 100,000 years our orbit gets million of miles closer to the Sun. The thing that is amazing about this is that recently the cooling and warming periods of Earth's current ice age come about every 100,000 years.

The energy from the Sun received on Earth varies with the square of the distance. This is science talk that means that a small change in distance means a much larger energy variation. So when you calculate time with orbit distance, you will find significant variations in the Sun's energy that gets to Earth.

Milankovitch Cycles

Milankovitch Cycles are named after Milutin Milankovitch, who combined orbit changes with two other types of Earth's movement.

Axial tilt is the inclination of the Earth's axis in relation to its plane of orbit around the Sun. Precession is the Earth's slow wobble as it spins on axis.

So Milankovitch combined orbit changes that occur every 100,000 years with the tilt that occurs every 41,000 years and wobble that occurs every 26,000 years. The combination is called Milankovitch Cycles.

These cycles cause periodic changes in Earth's climate. They are responsible for some climate change within their 100,000 year cycles. However, it is important to understand that Milankovitch Cycles do not cause major climate changes – such as million year ice ages. They are however important in determining which parts of Earth have more cooling or more warming

The major climate change comes from the variable amount of solar energy reaching Earth.

Chapter Summary

It is very possible that change in the orbit of Earth accounts for the cooling and warming periods of our current ice age.

The other planets in our solar system affect the orbit of Earth and our distance from the Sun. Orbit changes in combination with predicted solar changes could cause to start a long cooling cycle or a long warming cycle.

Earth's tilt and wobble account for climate change in various parts of the Earth.

Ninth Fact

The majority of scientists do not believe in man-made global warming.

You read that right. The climate alarmists wants us to believe that almost all scientists believe in man-made global warming, but that is not true.

"I do not know a single geologist who believes that (global warming) is a man-made phenomenon." – Peter Sciaky, Geologist

The truth is the vast majority of climate scientists believe catastrophic man-made global warming is pseudo-science. You've probably heard that over 97% of climate scientists believe in global warming. But that is a lie and the opposite is true. The vast majority of climate scientists don't believe in man-made global warming.

The 97% Myth
The Wall Street Journal wrote an article "The Myth of the Climate Change 97 percent". It is about the hyped up statement that climate change is man-made. Politicians such as Obama, Biden, Gore, and Kerry regularly claim that 97 percent of scientists agree that climate change is man-made.

64

> "I do not know of a single TV meteorologist who buys into the man-made global warming hype." – James Spann, meteorologist

We all know that it is very easy to lie with statistics. The survey results can be manipulated to prove almost anything we choose. And if you survey scientists who are not specialists in solar science, space science, geology, cosmology, physics, meteorology, astronomy and a few others, you answers are meaningless. The Wall Street article goes into some detail about the questionable methods used.

The statistic of 97% comes from a survey by Peter Doran and Maggie Kendall Zimmerman of the University of Illinois.

> "The public has been repeatedly misled that there is a scientific consensus on global warming. Totally false. Unfortunately, man-made climate change has become a political issue rather than a scientific one." – Glenn Speck, chemist, Environmental Lab

They polled 10,257 Earth scientists and 3,146 responded. Out of those 3,146 they handpicked a group of just 79 individuals who had previously given the "correct" answer to their climate questions. From that they got the results of 97% believing in man-made global warming. Those scientists who might not have agreed and voted against man-made global warming were specifically excluded. The results were very widely reported and are often used as supporting "evidence" that man-made carbon dioxide can warm the earth.

When you only give the survey to a very small group of people who already support your viewpoint, it is easy to get 97 percent of the people to agree with you.

Unfortunately, such shameful practice sets a dangerous precedent in a scientific field where open-mindedness is essential.

> "Let's be clear: the work of science has nothing whatever to do with consensus (which) is the business of politics. What is relevant is reproducible results. The greatest scientists in history are great precisely because they broke with the consensus." - Timothy Minnich, Atmospheric Scientist.

The vast majority of American top scientists believe man-made carbon dioxide is not a significant factor in climate change. To refute the 97% lie, John Coleman, co-founder of The Weather Channel, and over **thirty one thousand scientists** signed a petition stating the opposite position. They said that, "There is no convincing scientific evidence that human release of carbon dioxide gases is causing or will, in the foreseeable future, cause catastrophic heating of the earth's atmosphere and disruption of the earth's climate." Thirty one thousand is a lot of scientists but ignoring the facts of real science, political junk science still says it is a problem.

> "Global Warming: It is a hoax. It is bad science. It is highjacking public policy. It is the greatest scam in history." - John Coleman, co-founder of The Weather Channel

There is no basis for the claim that 97 percent of real scientists believe that humans are responsible for changing the climate and that it is a dangerous problem. Of course if you ask scientists who work for the government or get government grants, you could get 97 percent to agree on any talking point. I bet you could even get 97 percent to agree that the Sun goes around the Earth. Of course, the real issue is that 97% of scientists do not support man-made global warming.

> "By denying scientific principles, one may maintain any paradox". – Galileo Galilei, who was arrested for saying that earth revolves around the Sun that the Sun does not revolve around the earth.

Global warming became a cult following when Al Gore came up with the video named *The Inconvenient Truth*. With no debate, Gore's views magically became fact. Of course, Gore has a carbon tax "solution" which just happens to put millions of dollars into Gore's pocket because he owns a carbon credits company. Why won't Al Gore and other climate alarmists debate the issue? The reason they say the science is settled and will not engage in the debate is because they are wrong, and debating would prove them so. They just claim to have the facts and anyone who disagrees is a stupid denier.

> "Scientists all over this world say that the idea of human-induced global climate change is one of the greatest hoaxes perpetrated out of the scientific community. It is a hoax. There is no scientific consensus." – Dr. Paul Broun, U.S. Representative

In July of 2014 the ninth International Conference on Climate Change took place. Thousands of scientists attended the conference and voted that climate change was not a crisis and that climate change was far from settled. It was a dramatic demonstration of courageous people that came together to state that "global warming" was a complete hoax.

> "Being a scientist means being a skeptic." – Gerrit van der Lingen, scientist

Do not forget that Greenland was a lot warmer just a few hundred years ago when Eric the Red raised sheep and goats on its coastal grasslands. That's why they called it green. And now it is covered in ice. Earth has been cooling for the last eighteen years and there is nothing to indicate that this trend will stop anytime soon.

> "We have no reason to think that climate change is harmful if you look at the world as a whole. Most places, in fact, are better off being warmer than being colder. And historically, the really bad times for the environment and for people have been the cold periods rather than the warm periods." – Freeman Dyson, physicist

In the early 1970's, the climate alarmists claimed the world would run out of oil within 30 years. That's what science was telling us at the time, and a couple of gas crises later reinforced the message.

Of course, this turned out to be one hundred percent wrong. American technology used better methods to find and extract oil. As a result, hundreds of years' worth of more oil was discovered. As I write this, gas prices have gone way down and America is experiencing an unbelievable energy boom.

"The underlying effort here seems to be to use global warming as an excuse to cut down the use of energy," he said. "It's very simple: if you cut back the use of energy, then you cut back economic growth. And believe it or not, there are people in the world who believe we have gone too far in economic growth." - Fred Singer, Emeritus professor of environmental science

Chapter Summary

The vast majority of American top scientists believe man-made carbon dioxide is not a significant factor in climate change.

"The public has been repeatedly misled that there is a scientific consensus on global warming. Totally false." – Glenn Speck, chemist, Isotek Environmental Lab

Only 79 handpicked individuals, who had previously given the indication they believed in man-made global warming, were given the questionaire. From that they got the results of 97% of scientists believing in man-made global warming.

Tenth Fact

Climate change has become a pseudo-religion requiring no proof, just belief.

For some people, global warming has become a pseudo-religion. These people believe that they are saving the planet. Being green and reducing their carbon footprint is an Earth-worshiping religion. It is not science, because what sets science apart from religion is that in science nothing is expected to be taken on faith.

It's truly amazing how strong people's opinions are about global warming. Most of them know nothing about the science of global warming but often are absolutely positive that it is the fault of humans and carbon dioxide.

The eco-theology religion, like most religions, is authoritative with no room for other beliefs. Like other religions, the eco-theology religion's views on man-made carbon emissions are accepted as an article of faith. Almost no alarmists actually have the scientific background to determine for themselves if their beliefs in global warming are accurate. Like religion, alarmists are members of a faith-based organization.

> "I am a skeptic…Global warming has become a new religion." – Ivar Giaever, Nobel Prize Winner for Physics.

To these true believers, global warming seems to give their life meaning. This is similar to a religion where following God's word gives meaning to a believer's life.

Various alarmists have demonstrated that the global warming campaign is a religion. And worse, it is often an intolerant eco-theology religion. Case in point is the following quote from an article that suggests we should begin having Nuremberg-style war crime trials for global warming deniers.

> "When we've finally gotten serious about global warming, when the impacts are really hitting us and we're in a full worldwide scramble to minimize the damage, we should have war crimes trials for these bastards - some sort of climate Nuremberg." – David Roberts, Grist Magazine staff writer

Only someone with intense religious furor could equate differences in scientific viewpoints with the crimes of the NAZI Holocaust-style Nuremberg trials. Roberts sounds more like a jihadist ISIS Muslim who chops off the head of anyone who is not a true believer.

And to make matters worse, this eco-religion is mostly paid for with our tax money. With global climate change, there is no separation of church and state.

> "The global warming time bomb, disastrous climate changes that spiral dynamically out of humanity's control. These are the words of an apocalyptic prophet, not a rational scientist." – Dr. Nicholas Drapela, Chemist

Instead of a religious fast that restricts food, some alarmists have come up with a "carbon fast." Some groups send daily messages directing people to go on a "carbon fast." They encourage their followers to lower their "carbon footprint" and thus be one with the climate.

> "Global warming is a new religion and blasphemy against that religion is not a laughing matter. There is a great gap in Europe with the decline of any real belief in Marxism and any real belief in Christianity. This has filled the vacuum." - Nigel Lawson, former British Chancellor of the Exchequer

"True believers" tell us we should spend trillions of dollars to control a naturally occurring gas that has almost no effect on the climate. They have invented a fake carbon market to control this gas so that we can all feel we are good people. If you deny their request, you are evil or irrational or both. And most climate change true believers actually think they are doing good. Carbon religion is an amazing thing.

Even though he is a lawyer and not a scientist, Al Gore has emerged as the spiritual leader of this religion. He asks us not only to reduce our "carbon footprint" but also to use civil disobedience to force other people to follow his advice.

The Washington Times wrote that Al Gore's academic performance was rather dismal, particularly in the field of science. As a Harvard sophomore, Al Gore received a D in Natural Sciences. He deserves an F in global warming.

> "I believe we have reached the stage where it is time for civil disobedience to prevent the construction of new coal plants that do not have carbon capture and sequestration." - Al Gore.

Civil disobedience, really! The trouble with Al Gore as a spokesman for this alarmist global views is that he does not follow his own advice. He sees himself as so special, that he does not have to conserve carbon but instead wastes it. This is more than just Al Gore's hypocrisy; it's a question of his credibility. He flies around the world in private jets that use much more gas per passenger than commercial jets. Also his combined electricity and natural gas usage for his homes is about twenty times greater than the average American home.

When confronted with this inconvenient truth, Gore says that he buys carbon credits to offset his usage. But he neglected to mention that he buys the credits from the company he owns that sells carbon credits. The bottom line is he extravagantly wastes while telling others to conserve. When this became public, Gore was exposed as a hypocrite.

There is something wrong when wealthy, sanctimonious hypocrites want to imprison poor and middle class Americans into an energy-impoverished state. At the same time, they continue to flaunt their selfish, excessive carbon dioxide lifestyles.

"It is amazing to me, as a professional geologist, how many otherwise intelligent people have, as some may say, 'drunk the Al Gore Kool-Aid' concerning global climate change." – Earl Titcomb Jr., geologist

Chapter Summary

"Is climate change pseudoscience? If I'm going to answer the question, the answer is: absolutely...Global warming has become a new religion." – Ivar Giaever, Nobel Prize winner in physics

In recent times, there has never been another scientific theory where people who did not support the theory were so viciously attacked. This tells us that the theory is less about science and more about the politics or religious dogmas of the time.

Eleventh Fact

Government scientists falsified climate data in climate gate.

Climate gate is the revelation that an intergovernmental panel of scientists falsified much of climate data to convince the public of a need for cap and trade. The evidence shows outright fabrication of data. This is the biggest science scandal in hundreds of years. The climate scientists conspired to fiddle with temperature data. They attempted to prove their computer models were correct and to overstate the case for human influence on climate change. How can you believe anything from people already caught manipulating the actual numbers?

> "Not surprisingly, these (climate) models have been consistently and spectacularly wrong in their predictions, and always, amazingly, in the same direction." – Richard McNider, Professor of Climate Science and Dr. John Christy, atmospheric science

They attempted to stand climate history on its head by showing that, after many years of decline, global temperatures have spiked to their highest level in recent history. But soon thousands of leaked emails were released showing their dishonesty. The emails depict scientists discussing ways to subvert the normal scientific peer review process.

Over one thousand emails from top "man-made global warming" scientists revealed unscientific data manipulation. There was also evidence that data was manipulated. It was a conspiracy to destroy and replace the actual recorded information.

This group of scientists on the Intergovernmental Panel on Climate Change committee was caught cooking the books. Their emails showed their desperation at how difficult it was to get the desired results. So they decided to change the past temperature data from a basically flat chart into one which showed temperatures steadily getting hotter. Phrases such as "hide the decline" and "Mike's Nature trick" were discovered in their email correspondence.

They used this false data to promote a fake picture of world temperatures. It was called the "hockey stick" graph because of the way the graph looked. A "hockey stick" is a type of chart that would show a sudden higher temperature variance. Their goal was to further the arguments of the alarmists that there was a sudden increase in temperature. They called this "unprecedented modern global warming".

The "hockey stick" chart tried to show that temperatures in the Northern Hemisphere remained basically stable over 900 years, then suddenly spiked upward in the 20th century. This was presumably due to human carbon dioxide activity.

This fake evidence was supposed to prove the completely spurious conclusion of man made global warming and that the earth is suddenly heating up and is doomed. In fact it was the dishonest science that was doomed. All they proved is that political science is more important to them than real science.

The real numbers show that without their lies, there is no evidence of anything except normal temperature fluctuations within our current eleven thousand year old warming period. The fact that scientists would make up numbers to prove something that is not true is just plain wrong.

They used this compromised data for their predictions that the world would heat up to catastrophic levels unless trillions of dollars are spent to avert it. Their goal was to stifle scientific debate on global warming and win new government grants, money, and programs to fight their imaginary pending disaster. If you can't trust these "man-made global warming" scientists not to change the data, then you can't trust anything they say.

> "The questions are scientific, but the UN answers are political. The global warming debate is hardly about science." – Allen Simmons, Computer Modeler and Engineer

Through their scare tactics they hope to transfer many billions of dollars from the average hardworking American and give it to their corrupt political friends. It's called cap and trade, and is really a tax on electricity, heat, manufacturing and transportation fuel that will be passed along to average Americans as higher prices on almost everything. The way it works is that every business and ultimately every person will be taxed on their "carbon footprint" and the money and credits will be given to Al Gore type groups who will no doubt keep much of it. This is wrong, and would be the largest tax increase in American history, putting money in the pockets of the politically connected.

"In recent weeks I have been the target of attacks in the press by various radical environmental and politically motivated groups. This effort should be seen for what it is: a shameless attempt to silence my scientific research and writings, and to make an example out of me as a warning to any other researcher who may dare question in the slightest their fervently held orthodoxy of anthropogenic global warming." – Dr. Willie Soon, astrophysicist and a geoscientist

Since most people no longer believe in man-made global warming, legislation on carbon dioxide will be wrapped up in the language of energy security, job-creation, and boosting American competitiveness, rather than by talking about the climate changing, about which voters appear relatively unconcerned. Of course the real results would be job destruction, high inflation, more dependence on the terrorist countries, and a less competitive America.

"The recent 'panic' to control greenhouse gas emissions and billions of dollars being dedicated for the task has me deeply concerned that US, and other countries are spending precious global funds to stop global warming, when it is primarily being driven by natural forcing mechanisms." – Dr. Diane Douglas, Climatologist

If you do not believe that human civilization is the cause of climate change, the media defines you as a "denier." As a denier you are by their definition naïve or worse. But in addition you're also at odds with science and with "overwhelming" scientific evidence. Except their evidence is fake and very underwhelming.

> "The whole hockey-stick episode reminds me of the motto of Orwell's Ministry of Information in the novel 1984: 'He who controls the present, controls the past. He who controls the past, controls the future." – Dr. Will Happer, Princeton University Physicist

Below is a real chart of the average temperatures during the last 16 years. The temperature change (in degrees F) is shown on the right. Note that without manipulated climate gate numbers there are no indications of global warming. The chart only shows normal variations that would be expected if there were no global warming.

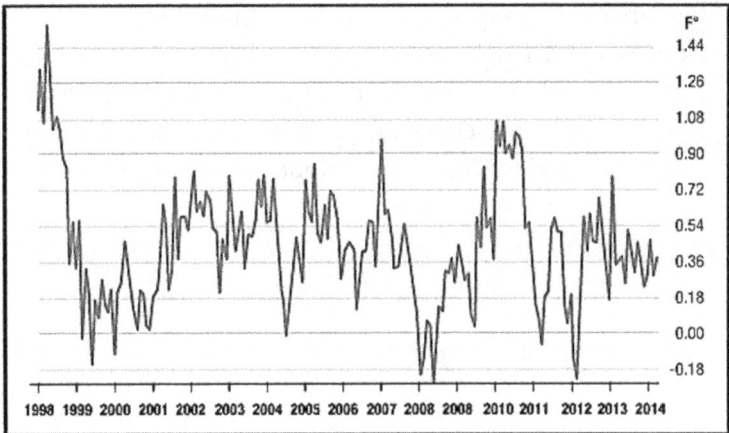

Apparently climate records are still being manipulated. On June 23, 2014 the Washington Times reported evidence that shows the raw American temperature records from the Energy Department were still being altered.

80

Additionally, many alarmists still continue to use the manipulated data as if it were true. Tell a lie often enough and people will believe you.

> "We've seen temperatures hold steady for the past 18 years, as measured by Remote Sensing Systems, the most accurate thermometers in the world." – Washington Times

Chapter Summary

Climate Gate proves that man-made global warming is a lie. Because of the incrementing emails, the hockey stick theory was shattered. It is a scam by greedy government officials and dishonest scientists. It is about money and power. They want to take your money and your power.

> Science is rarely determined or finalized; science evolves and the huge complexity of climate science will certainly continue to evolve in the light of new facts, new experiences and new understandings." – Geoffrey Kearsley, geographer, environmental communication

Twelfth Fact

Money and power can cause people and the news media to promote global warming.

"In one of the more expensive ironies of history, the expenditure of more than $50 billion on research into global warming has failed to demonstrate any human-caused climate trend, let alone a dangerous one." – Bob Carter Paleo-climate scientist

The government currently wastes about twenty-two billion dollars of America's borrowed money every year on global warming. If we spent this on securing our borders we could easily stop most criminals and terrorists from illegally crossing into America.

In addition, more than one trillion dollars a year is wasted on unnecessary government requirements that result in higher costs to working people. It is estimated that the average cost to an American family will be about $4,000 every year. And that will be rising. Worse still, some three million Americans have or will lose their jobs because of these government regulations.

And to top it all off, the cap and trade tax and all those government regulations would have no discernible effect on either carbon levels or the constantly changing climate of our earth.

"If, back in the mid-1990s, we knew what we know today about climate, Kyoto would almost certainly not exist, because we would have concluded it was not necessary."- Dr. Tad Murty, Climate researcher

Cap and Trade

Most non-government scientists agree that the Sun causes global warming and cooling. Water vapor and clouds play a role. And to a much lesser extent, carbon dioxide plays a very minor role. Only political science still claims that man-made carbon dioxide is the determinant factor of global warming.

So what is this cap and trade that the politicians are trying to put into law? Basically it is paying money to Al Gore and other elite cap and trade wealthy and connected people. It is really a tax on everyone that ends up in the pockets of politicians as bribes, pork barrel and just plain corruption.

> "Kyoto (global warming cap and trade treaties) is one of the most aggressive, intrusive, destructive ideologies since the collapse of communism and fascism." – Andre Illarianov, The Russian Academy of Sciences

Who pays for the cap and trade tax? We all do. The price of almost everything we need and use goes up.

It is obvious that the price of gasoline will go up. But that affects millions of products that are shipped from one location to another. All trucks and cars use fuel that would end up costing more.

But this is not a tax on the corporation that makes or transports the products. Corporations always need to make a profit and don't end up paying for taxes. They just raise the selling price so that all of us consumers end up paying the tax.

Heating and electrical costs are already higher due to government restrictions on coal. They could continue to increase greatly since most of the coming cap and trade tax would just be passed along to all of us consumers. It would raise the cost to heat or air condition our homes and businesses as well as making us pay more for our transportation and our manufacturing. Almost everything would go up in price.

> "The data which is used to date for making the conclusions and predictions on global warming are so rough and primitive, compared to what's needed, and so unreliable that they are not even worth mentioning by respectful scientists." – Dr. Gregory Moore Aerospace and Mechanical Engineer

Who ends up with the money? The wealthy blue blood one percent of the people such as Al Gore have already set up businesses to make sure they get a piece of the action. They don't pay the tax. Because of them, the rest of us would effectively get poorer. They are more devious than you could have ever thought.

> "The American people are fed up with media for promoting the idea that former Vice President Al Gore represents the scientific consensus." – U. S. Senator James Inhofe, Chairman of the Environment and Public Works Committee

Al Gore is one of the one per centers but who gets hurt the most? The other ninety-nine percent of working and poor people pay the price. They're the ones that don't have any optional dollars to spend. It will also hurt all the people involved in the coal industry since they did or will lose their jobs and end up on food stamps or much worse.

Cap and trade has nothing to do with global warming. It has everything to do with redistribution of income, government control of the economy, and a criminal payoff to dishonest politicians.

Stupid government mandates

Ethanol in our cars is just one example of government wrong thinking about climate change. Ethanol is pure alcohol like vodka except in a purer form. Instead of being 80 or 100 proof, ethanol is 200 proof.

The normal thing to do with corn is to eat it or feed it to chickens or other animals and eat them. But now much of that corn is used to make ethanol. Then we mix it with gasoline and put it in automobile gas tanks. So, in effect, we burn the corn.

Ethanol costs more per gallon than gasoline and it provides worse mileage than normal gasoline. But it makes climate alarmists feel good and some people are making a lot of money in the manufacturing process.

Taking into account the energy used in make and use fertilizer, planting, growing, harvesting, distilling, and shipping, ethanol produces more carbon dioxide than gasoline. Additionally, ethanol corrodes older car engines, so it is responsible for more air pollution than gasoline.

Most importantly, it uses up food that could feed the hungry. So there is less food to eat which results in raising the cost of food for everyone. Just one large tank of ethanol gas burns enough corn to feed a starving person for a year.

High food prices and shortages were one of the real causes of the Arab spring that occurred some years ago.

The food that remains after meeting government requirements for food burning isn't enough to supply the meat and milk production operations at a reasonable price. Consequently, many meat and milk production operations must raise their prices or close their doors because of the high cost of animal feed grains.

So why do we do such a stupid thing? The only reason we do it is to promote one world government where the political elite can get and retain money and power without having to go through the election process.

Most people who have the facts think ethanol in gasoline is stupid and downright thievery. Hungry people think it is criminal. Thank you Al Gore!

> "The sky is not burning, and to claim that it is amounts to journalistic malpractice...The press only promotes the global warming alarmists and ignores or minimizes those of us who are skeptical."
> – Dr. Mark Campbell, chemistry professor

Chapter Summary

The cap and trade tax would hurt poor people the most. They're the ones that don't have any optional dollars to spend. It will also hurt all the people involved in the coal industry since they will lose their jobs and end up on food stamps.

Cap and trade carbon taxes are simply a fraud by groups of greedy people and the rich blue bloods to scam money from the working people of America. They would not do any good and even worse they would be very harmful.

Thirteenth Fact

Political lies and manipulation cause the global warming problem

Big government politicians want as much power and control as possible. They already have taken over the banks and health care. Now they want to take over energy. Almost everyone uses energy so this would be a major coup and result in significant control of everyone in America. Climate change is the ideal scare tactic for big government to drastically increasing their control over our country.

> "Unfortunately the global warming hysteria, as I see it, is driven by politics more than by science... The idea that global warming is the most important problem facing the world is total nonsense and is doing a lot of harm" – Freeman Dyson, a brilliant scientist who thought alongside Einstein, Robert Oppenheimer, Richard Feynman, Niels Bohr and Enrico Fermi.

Why do so many people believe in man made global warming? Most people are not scientists and just want to do the right thing. They believe they are saving the planet, and that of course is the right thing to do. They do not realize that man-made global warming is not fact and is just the result of political manipulation of their minds. They do not realize that Mother Nature and not humans control Earth's temperature.

The manipulative (brain washing) actions of the global warming alarmists coincide perfectly with a book titled *Rules for Radicals* written by Saul Alinsky. He was a

Marxist anti-establishment activist. His book described concrete tactics to lie, manipulate, and use dishonesty to fool the voters and get the desired results. Alinsky's advice was to do and say whatever it takes to gain power.

Alinsky taught the dark art of destroying political adversaries. He showed the radical socialists how to achieve their agenda by any means necessary. He promoted harassing those with opposing viewpoints and to deepen existing divisions existing in the population. This means turning blacks against whites, rich against poor, etc. The goal of his book was to change the current view of his opposition and move you closer to his desired viewpoint.

He advocated that the end justifies the means regardless of the inconceivable damage to some people. He asserted that to promote change, people must feel defeated, frustrated, and hopeless. Obviously, people would then need the hope and change that the socialists will provide them. And that change is whatever they desire.

The basic concept of Alinsky is to overload the government bureaucracy with a flood of impossible demands, thus pushing society into crisis and economic collapse. Global warming and the flooding of America with thousands of illegal aliens are just two of the many manufactured disasters that America will unsuccessfully seek to resolve.

> "Power is not only what you have but what the enemy thinks you have" - Saul Alinsky

Paraphrased, this quote means to make a big enough noise so that people believe your group has the numbers on your side. This is exactly where the "97 percent of scientists" quote comes from. This is pure Alinsky politics. When you only give the survey to a very small group of people who already support your viewpoint, it is easy to get 97 percent of the people to agree with you. You could even get 100 percent, but of course that would more easily be seen as fake.

> "Whenever possible, go outside the experience of an opponent. Here you want to cause confusion, fear, and retreat... Keep the pressure on. Use different tactics and actions and use all events of the period for your purpose." - Saul Alinsky

Very few people know much about the science of climate change so it is certainly outside their experience. Also the alarmists are certainly keeping the pressure on. They use current events and blame carbon dioxide for all occurrences of terrible storms and devastation. It is almost like watching a disaster movie such as Armageddon or War of the Worlds.

Science is a very powerful and carries a lot of weight. However, some of these scientists lied and sold us out. With government funding of so many scientists, we need to know how and why their scientific decisions are made.

> "Ridicule is man's most potent weapon. It's hard to counterattack ridicule, and it infuriates the opposition, which then reacts to your advantage." - Saul Alinsky

There is an expression that says, "If you can't beat them with logic, ridicule them." This is exactly why they hung the name "deniers" on the scientists who told the truth about climate change. If they were not afraid of a real scientific discussion they would not need to use ridicule.

This of it self proves that human caused climate change is a complete lie.

They lump "deniers" with the crazies so they can avoid legitimate questions. That way, inconvenient facts such as the Sun, ice age cycles, and water vapor can be totally ignored.

Alinsky was a radical socialist and would now be called a progressive. He believed that a big powerful government was best for humans. But socialism and communism almost always turn into dictatorships and steal individual freedom.

Hitler is a case in point. He ascended to power with the National Socialist German Workers Party. In German National Socialist is spelled National Sozialistische. The first two letters of **Na**tional were combined with the third and fourth letters of So**zi**alistische which together spell NAZI.

> "It is easier to silence scientific dissent by utilizing the politics of personal destruction, than to actually debate them on the merits of their arguments. That should tell you something about the global warming debate." – Mike Thompson, Chief Meteorologist

Alinsky was a radical socialist. The ultimate goal of global warming is to change America and the world into a far left socialist style strong central government. This government would eradicate almost all of our freedoms and likely evolve into a dictatorship. This would all be done under the false umbrella of saving the Earth.

Global Cooling

> "It is one thing to impose drastic measures and harsh economic penalties when an environmental problem is clear-cut and severe....It is foolish to do so when the problem is largely hypothetical and not substantiated by observations...We do not currently have any convincing evidence or observations of significant climate change from other than natural causes." – Frederick Seitz, past president, U.S. National Academy of Sciences

So our climate is constantly changing, but is getting warmer or getting colder best for humanity? Well there are real problems with getting cold. One of the main problems is that food does not grow very well when it is cold. Much of the land where food is currently grown would no longer be productive. There would be severe food shortages and starvation would be common.

> "All those urging action to curb global warming need to take off the blinkers and give some thought to what we should do if we are facing global cooling instead." – Phil Chapman, astronautical engineer, NASA astronaut

Heating bills would be astronomical, and most of humanity would have to relocate to areas closer to the equator. This would likely cause wars and numerous other problems. Global cooling would definitely be a disaster.

But it would not be the end of the world. Humans would survive just as we did in the past when our ancestors lived in the caves, often without fire. Today we are in a much better position to survive than those ancient cavemen.

> "Billions of dollars of grant money are flowing into the pockets of those on the man-made global warming bandwagon. No man-made global warming, the money dries up." – James Spann, meteorologist

In comparison to cooling, global warming problems would be minor. Slightly more heat and carbon dioxide would increase food production. It would force some people to relocate away from the seashore, but far fewer than in the global cooling scenario.

The cooling periods of the ice ages are much worse times for humans. By contrast, the warming period of the last eleven thousand years increased the human population from about five million people to over seven billion people. It has also brought civilization to much of the world.

> "Governments are trying to achieve unanimity by stifling any scientist who disagrees. Einstein could not have got funding under the present system." – Nigel Calder

The problem with the man-made global warming theory is that since 1998, temperatures have been dropping. There has been no evidence of global warming for 17 years and counting. It is expected that Earth will continue to cool for at least the next 20 years, and possibly a lot longer.

In 2007 Al Gore said that the North Pole ice cap could be completely gone by 2014. Of course, it didn't happen, and the North Pole ice cap has grown about 50% since Gore's prediction.

Chapter Summary

Many of the methods used by the alarmists are just political rules for stealing the hearts and minds of people. Political science likes to say that discussions are over and their side wins. Real science always keeps an open mind to new theories that better explain physical reality. The alarmists are not about real science but instead they are about political science.

> "Observe which side resorts to the most vociferous name-calling and you are likely to have identified the side with the weaker argument and they know it." – Charles R. Anderson, Research Physicist

Global warming would cause some problems but they are nothing compared to huge problems with global cooling. One of the main problems is food does not grow very well when it is cold. Much of the land where food is currently grown would no longer be productive. There would be severe food shortages and starvation would be common. Humans have always done much better when the climate was warmer, and worse when it was cooler.

Conclusions

Earth has always had climate change and it will always continue to have climate change. Mother Nature and not man is responsible for any global warming or cooling.

Alarmists say that it is humans that are causing climate change. They say that it is the carbon dioxide that humans release that is making Earth warmer. They claim that Earth would be a disaster if everyone doesn't give them money and do exactly as they say. But they lie!

> "As the glaciological and tree ring evidence shows, climate change is a natural phenomenon that has occurred many times in the past, both with the magnitude as well as with the time rate of the temperature change that have occurred in the recent decades." – Gerhard Lobert, physicist

During an ice age there are numerous cooling and warming periods. We are currently in a warming cycle of Earth's fifth ice age. It has already lasted about 11,000 years and will likely end within the next thousand years. Earth has recently been getting colder, and we may be heading for a return to a cooling cycle or this may just be a colder period in our warming cycle.

Changes in these climate cycles are not your fault. They are just the results of the forces put into play by Mother Nature.

Almost all the heat on Earth comes from our Sun. So any change in the Sun's output would certainly affect Earth's average temperatures.

Over periods of thousands of years, the Sun itself is changing. At this point in time, the Sun is getting dimmer and less energetic. That means Earth is getting less heat. We are moving into a cooling period. It may last only twenty years, but it also might be the beginning of our expected move into the cooling portion of our current ice age, which could last as many as ninety thousand years.

> "As a scientist and life-long liberal Democrat, I find the constant regurgitation of the anecdotal, fear-mongering clap-trap about human-caused global warming to be a disservice to science." – Dr. Martin Hertzberg, meteorologist

The main greenhouse gas is water vapor. Water vapor accounts for about 95 percent of the "greenhouse" effect.

This information is entirely skipped by the alarmists because otherwise they would not get the results they want. Their computer models do not show water vapor even though it is by far the absolute number one most important climate change gas in the world.

> "Compared to solar magnetic fields, however, the carbon dioxide production has as much influence on climate as a flea has on the weight of an elephant." – Oliver Manuel, professor of nuclear chemistry

Carbon dioxide is not a poison or a bad gas. Carbon dioxide is not a pollutant or smog. Carbon dioxide is a plant food and is necessary for green plant growth and for our very lives. Earth would do better with more carbon dioxide and not less.

The role of man-made carbon dioxide in changing global temperature is insignificant. It is too sparse in the atmosphere and the band of infrared energy that it can absorb is to small.

The human contribution to carbon dioxide is only 5% of that 400 parts per million. Therefore, human contribution is only about 20 parts per million. This is so small it's not even enough for a serious discussion of it causing global warming. The alarmist's entire viewpoint is based on thin air.

> "Climate change is normal. There have always been phases of climate warming, many that even far exceeded the extent we see today... Because of climate's high complexity, reliable prognoses just aren't possible. Nature does what it wants, and not what the models present as prophesy. The entire carbon dioxide debate is nonsense." – Klaus Eckart Puls, physicist and meteorologist

Changes in atmospheric carbon dioxide follow changes in Earth's temperatures. As our oceans heat up more carbon dioxide is released into the atmosphere. A cause does not follow an effect. The opposite is true. So the cause of changing carbon dioxide is Earth's changing temperature.

Water exaggerates the changes in our climate. It can make the Earth either cooler or warmer.

> "Even if the concentration of 'greenhouse gases' double man would not perceive the temperature impact" – Oleg Sorochtin of the Institute of Oceanology

Carbon is almost everywhere. It is the fourth most abundant element in the universe. It is the second most abundant element in the human body. Furthermore, carbon is the chemical basis of all known life in the universe.

Science indicates that neither human activity nor carbon is the cause of global warming. And in fact, Mother Nature is the cause. Only a politician would think that we have to waste many trillions of dollars to try and reduce something they call our carbon footprint.

> "Scientists all over this world say that the idea of human-induced global climate change is one of the greatest hoaxes perpetrated out of the scientific community. It is a hoax. There is no scientific consensus." – Paul Broun

Politicians such as Al Gore use "global warming" as a scare tactic in order to influence us to buy cap and trade credits from them. It's working and they are already counting the money. But their Chicken Little "the sky is falling" routine will backfire on them. Already more and more people are seeing through their fire and brimstone routine.

Climate change alarmists don't like to be confused by the facts. They have their minds made up and the global warming religion fills their needs.

Today it can be seen that the alarmists' predictions are 100% wrong. Even by the best estimates, Earth is not hotter and they can't explain the missing heat energy. They predicted the temperature would increase but it didn't. They predicted the oceans would warm and they didn't. They predicted the polar ice would melt and it grew. The multi-billion dollar computer climate programs built by the alarmists have been wrong and incapable of predicting the future climate.

"Global warming as proclaimed by Al Gore and Company, is a hoax." – John Takeuchi, meteorologist

Various opinion polls indicate that most Americans rate climate change as unimportant compared to more immediate threats. Even with all the alarmist hype, they feel that climate change doesn't have an impact on their day-to-day lives. Things like Islamic terrorists, jobs, the economy, disease, crime, etc. were rated as some of their more important concerns.

In the Southern Hemisphere, Antarctica has been cooling and gaining ice for years, according to data gathered by the Department of Atmospheric Sciences at the University of Illinois.

Since 1958, ground-based thermometers show an average temperature rise of about 0.7 degrees Fahrenheit. However, more accurate temperature measurements made from satellites began in about 1975. They show no average warming trend.

There is no good scientific evidence that human activity is causing climate change. By contrast, there are reams of scientific evidence that Mother Nature causes climate change and has been doing it long before humans were on Earth.

Humans have been on Earth for only two hundred years. Our current ice age is already more than two million years old.

It was only about fifteen thousand years ago that humans learned to make their own fire. Before then they relied on natural fires, such as lighting strikes to start fires.

Climate change is a very complicated non-linear process. Many of the variables are still not understood. Important items such as sunspots, clouds and water vapor and other important forces are not even included in current computer models. Therefore, these models are completely unable to make an accurate prediction of Earth's climate.

> "I used to agree with these dramatic warnings of climate disaster…However, a few years ago, I decided to look more closely at the science and it astonished me. In fact there is no evidence of humans being the cause. There is, however, overwhelming evidence of natural causes such as changes in the output of the Sun." – Dr. Ian D. Clark, Paleoclimatologist

Still climate change alarmists have woven a fabric of distortion, scientific dishonesty and deception to push their view of man-made climate change. Certainly some of this is deliberate and calculated to make money or increase their power. But most comes from people who do not really understand the complicated science but just think they are doing something good to save the world. They often are well meaning people who really believe that if the world spends untold trillions of dollars we can be more powerful than Mother Nature and control Earth's climate. For some, the global warming concept has become almost a religion.

Ice ages and the warming and cooling cycles within them are predominantly the result of a complicated interaction between the changing heat from our Sun, water vapor the changing cloud cover and the changing distance of Earth from our Sun. Other things like volcanic eruptions and meteor strikes on Earth can contribute to the change.

Changes in the concentration of the trace gases of carbon dioxide are the results of climate change and not the original cause of it.

The greedy global warming farce is not just wrong because it is a lie. It is also dangerous. After crying wolf about Earth getting warmer, few people will believe that Earth is getting colder. And, if the evidence continues to accumulate that Earth is getting colder, we must begin planning for those eventual effects.

> "It's going to be dangerously cold, as certainly as it was in 1793-1830. Right after the formation of our country, we had a disastrous cold climate on the planet and it's coming back around." – John Casey, climatologist

Our Sun's activity is getting weaker. Its recovery from the last sunspot minimum took longer than before. Consequently, unless there is a change in the Sun, Earth will continue to grow colder. But it is still possible that the Sun's activity will begin to get stronger and warm Earth more. The bottom line is that no one knows for sure, but the odds are that Earth will get colder and the oceans will get lower.

And Earth is already starting to get colder. For example, the North Polar ice cap is actually increasing in size. Recent satellite images from NASA show an increase of about 50 percent.

> "Warming fears are the worst scientific scandal in the history...When people come to know what the truth is, they will feel deceived by science and scientists." Kiminori Itoh, environmental physical chemist

As Earth gets colder, the water vapor in the air decreases, producing expanding drought conditions and reduced rainfall. That, plus colder temperatures, will reduce crop harvests. The jet stream will shift and possibly result in more droughts in the southwest and more flooding in the east.

So the needed planning would be ways to relocate our agriculture and save it from the coming cold. It may be enough to relocate our agriculture to the southeastern areas that are not as effected by the global cooling. Also large-scale (real) greenhouse operations in the permanently cold areas will help. However, if this is really the beginning of an actual cooling period of our current ice age, additional plans need to be made.

Genetically modified foods that resist cold and drought would probably be created. Much agriculture may eventually have to be moved further south, perhaps near the equator. But it will not be the end. Humans survived through the last cooling cycle of our present ice age and they will survive now. And we will have plenty of advance warning to plan for our future.

> "Gore's vision of a world plagued by rising seas and rampant tropical diseases, which he preaches can be averted only if Americans drive hybrid cars and replace energy-guzzling dryers with outdoor clothes lines, is as false as the notion that the ice sheet is melting in the North Pole, threatening to drown frisky polar bears." – New York Post 2014

1. For millions of years Earth has been repeatedly cooling and warming. In its extremes it changes between ice ages and warm tropical weather.

2. Earth's heat comes from the Sun and there are continual changes in the amount or energy the Sun sends to Earth.

3. The main greenhouse gas is water vapor. In comparison to water vapor, man-made carbon dioxide's greenhouse effects contribution is insignificant.

4. We know that human contribution to the total carbon dioxide in the atmosphere is very small.

5. Cap and trade will not change the climate but it will make certain politicians and their friends very rich. It will also ruin our economy by raising prices on almost everything we buy or use.

6. The climate gate scandal proved that many government paid scientists manipulated data to make things seem much worse than they really are.

> "The longer trends tell us that by 2020, we will be experiencing an unusually low-energy Sun. Apparently, these are exactly the conditions that preceded the Maunder Minimum and ushered in the Little Ice Age." - Geoffrey Kearsley, geographer, environmental communication

The belief that we can be so certain of climate temperature 100 years from now when we don't know what the temperature will be one month from now is ridiculous. The belief that burning food to make ethanol and damaging our economy will make our temperatures cooler is insane. That we would even consider ruining our economy and paying money for carbon credits to a money grabbing, controlling political elite is just mind boggling.

No one knows for sure if it will be colder or hotter for the next thirty years. But judging from the reduced solar activity the best bet is that it will get colder. Perhaps we need to change the slogan "Stop global warming" to "Stop global cooling."

> "If you look at the 100-year global temperature chart, you look at the steep drop-off we've had since 2007. It's the steepest drop in global temperatures in the last hundred years...There is no human on earth, much less here in the U.S., who has experienced the depth and duration of cold we're about to experience — it's that serious." - John L. Casey, climatologist and former NASA consultant

Nobody has ever been able to prove that climate change is man-made. It is just a theory, and a false one.

About The Author

The Sputnik period of 1957 marked the success of the Soviet space program and the beginning of the space race. Still in high school, Alan Fensin was already interested in space and put together a team that built and tested small rockets.

After college, Alan began his career with Boeing and NASA in the early days of the American space program. He was a key member of the Apollo rocket design team that successfully put a man on the moon. As an electrical engineer, Alan helped design many of the critical elements used in the electrical system of the Saturn 5 moon-rocket.

To this day he continues to be interested, and to keep up with all things concerning space, our solar system, and the workings of our climate.

He has been a lecturer and writer for the last twenty years.

www.ingramcontent.com/pod-product-compliance
Lightning Source LLC
Chambersburg PA
CBHW050541280326
41933CB00011B/1672